AN ENGINEERING

VIEW

OF THE UNIVERSE

VOLII -

THE SOLUTION FOR PI

By Robert Heilman

TABLE OF CONTENTS

COPYRIGHT PAGE

THE BEGINNINGS

Never thought I would write a book, especially about the state of Physics and the Universe, But, you never know. My name is Robert Heilman. When I graduated from High School I actually thought I was going to be an Accountant. So off to College I went and began learning Accounting. I was introduced to computers that were being integrated into the Business world. I learned Programming and computer code was like a breath of fresh air compared to Accounting. I loved the creativity and clear output results. To support myself while attending school, I landed a job at a local automotive parts manufacturer. Because Computers were starting to lead robotics Into manufacturing, I found myself being pulled into Programming more and more. Getting my hands dirty, creativity, working with numbers, precision, and measureable results, I thought this was IT. But, as things go, the company was bought out and the staff reduced. I found myself without a job. So I began looking, But I was convinced that Accounting was out. The type of jobs I was looking for were far and few between, especially in my small Town. 60 miles away, in a city called Detroit, jobs were plentiful. Because of my computer and programming experience. I landed a job as an NC programmer and Design

Studio support for an Automotive supplier. I was exposed to many things I never knew existed! Computer Aided Design, Computer Aided Simulations, Crash worthiness, weight reduction, Aerodynamics and wind tunnel testing, Strength of materials, Road Testing, In Process Testing, Analyzing Reports, Writing Reports, and many others; quality, design, production, inspection, cost reduction methodologies from the US, Japan, Germany and around the world. Things such as Statistical Process Control, Six Sigma, Etc. This was Engineering! It was a natural fit, especially with the integration Of computers. Within a few years I was an Engineer And never looked back.

ON TO PHYSICS

Why this series of books? Why now? Well, it would be easy to say I am of German ancestry and Germans have a long history in Physics. I MUST do this book, because cousin Einstein needs a hand, or some other nonsense. But the biggest driver is that I thought that using Engineering logic, I could help find Dark Matter/ Dark Energy. I started researching and reading Loads of articles, most by Physicists, some by Anybody. I was just searching for the truth. As I researched, questions kept popping up that didn't seem right from an Engineer's point of view. So what does that mean? Engineers are physical science people: is it real? What are the mechanics: what does it do? How does it do it? What is the effect of what it does? How can it be made better? And then the actual Physical properties questions; what does it weigh? What are the dimensions? what is it made of? And finally. Test, Test, Test. Sometimes we run tests Just to make things Fail, so we know it's capability and reliability. Most of the tests are video taped and we review and analyze. Some tests, especially Safety tests, are ran by independent Test Labs, with NO stake in the outcome and only care about running the test properly. So, people can tell much better Stories than I tell, and can Out-Calculate me all day long, but where the rubber meets the road in Physical Testing, I will Own you. After a few thousand tests, you develop instincts for how things Mechanically work and interact. And finally, when I read a Theory, or about a test, or even just the movement and interaction of objects, this jumps out at me like it's in bold print. This Book addresses the questions in the Universe as seen by an Engineer.

Wanted to do a recap of Volume I here but I won't make you wait:

SOLVING FOR PI

Mille and Zoe, my braintrust (actually Puppies) occasionally look at me quizzically like they don't understand what I'm talking about. Such is the case for PI. Other people have asked me, actually challenged me, about this. I have been putting it off as its not as much fun as The Universe, and math gives me a headache. Ok, so here it is, Occam's Razor, simple answer. THIS IS NOT A MATH PROBLEM. This is more of a Geometry problem. When you use math, it needs to be used to explain what is happening in a physical Universe, In other words, Math is not a Solution, it's an explanation.

PI = 3.15

Why would an Engineer, writing a book on the Universe, include a Chapter on Pi? One could say that the Universe includes all, so anything is fair game. But the fact is that I notice something wrong; logic, proof, conclusions, etc. and I must respond (the German in me). The truth is this should be so simple I don't know what went wrong for hundreds of years. What do I mean? Pi=circumference/diameter, pencils down, computers off, end of story. For an Engineer this is right up our alley; measure the circumference, measure the diameter, divide, repeat that a few thousand times, then ask for a raise.

Somewhere things went horribly wrong. Engineers did not get involved. Maybe they thought it was so simple even a Physicist, oh sorry, I meant Mathematician, could do it. Whatever, where is the data? Hundreds of years and no data, only calculations. I get it, our precision for making and measuring a perfect circle is only good to 3 or 4 decimal places, but it should still exist even as a directional indication.

What Is Wrong! Let's fix this; even though I prefer physical data, we can break this down. First a little History: The ancient Greeks; Euclid, Archimedes, Pythagoras, et al, gave us most all of the formulas for triangles. Why is this important? Because the quest for Pi has centered around multi-sided polygons and triangles to approximate Pi. It is believed that if a circle can be divided into smaller and smaller triangles(billionths of a degree), the straight line bases of the triangles will at some point match the curved circumference of a circle. But as one University of Michigan student told me; a straight line will never equal a curved line no matter how small the pieces. And this leads to an irrational quest for Pi. Ok, back to the Greeks and the formulas.

For some unknown reason the Greeks included irrational numbers in their formulas. What do I mean? Let's take the Pythagoras Theorem. In a right triangle, the side adjacent squared plus the side opposite squared equals the hypotenuse squared. SO, if side adjacent is 1 and side opposite is 1, the hypotenuse is equal to the square root of 2 which is an irrational number. This is Absolutely wrong! Why? Because these are not just numbers they can be thought of as measurements. Side opposite has an exact measurement of 1 and side adjacent also has an exact measurement of 1 and since the hypotenuse is connected dot to dot from one end of the side adjacent to the end of side opposite it too has a define length. An irrational number is not a length and therefore cannot be a side of a triangle. The Greeks are somewhat wrong in the triangle formulas. Now let's get back to Pi.

Start with the knowns or truths. We know Circumference/Dimeter= Pi. We also know that both the Circumference and the Diameter can be measured from each start point to an end point and therefore are not endless strings, but a fixed length. Wait you say, Pi is not a measurement, it is a ratio and that could be irrational. Nice try Grasshopper but since Diameter can be 1, 10,100 or multiples of 10, in this case Pi would be a fixed Circumference length divided by 1; a fixed number, and not irrational. Yes, 2 through 9 could all still be irrational, but the idea of a fixed ratio for Pi would be disproved. UNLESS, Pi is evenly divisible by 2 through 9 (chart below). Then further, current claims to Pi or any irrational number are WRONG!

I know this hurts, but want more Truth? Okay, a picture:

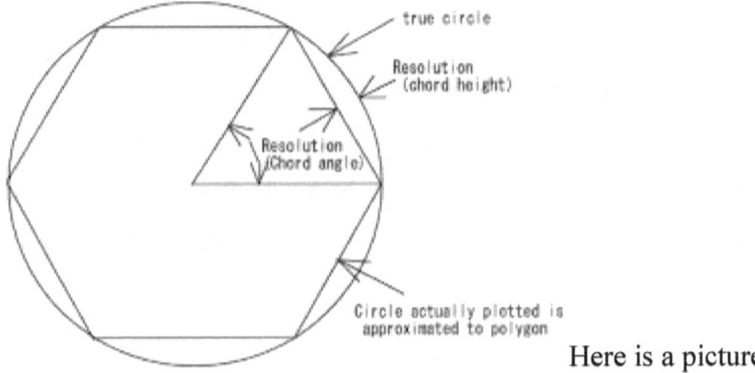

Here is a picture of a multisided Polygon compared to a circle. Notice how much bigger the circumference of the circle looks to the Polygon. But for a 360 faced polygon, the height difference from a straight line to a curved circle for any face is approx. .0004! Easy

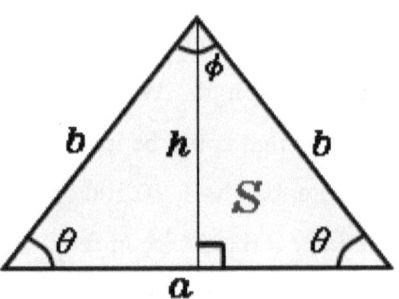

to see why Archimedes thought that 360 degrees made a circle. Keisan Casio pic.

But that never seemed right to me; I investigated. With the help of Ke!san Casio Online Calculators, which allow 50 digits for triangles and 102 for general math. I found a small problem with the formulas: The circumference of a polygon can never equal the circular circumference or as my University of Michigan son told me, no matter how small a straight line will never equal a curved line, so why look at straight line solutions? The proof of this is that if we take a 360 degree circle and look at a 1 degree slice or a triangle (one face of a 360 face polygon). This 1 degree slice is an Isosceles triangle with equal sides, with each the length of the radius(b); a bisecting line(h) can be generated but the length will never equal the radius lines, even if the angle at the top is so small that the 3 lines (b-h-b) almost touch each other. What happens is that if you divide a circle into a billion triangles, each height(h) will be short of the radius by a billionth, and a trillion by a trillionth and so on. This not only creates an incorrect irrational condition but is just WRONG, because we already know a circumference has a fixed length. Even if that length is to a billion of inch, there is an end to a measurement. At this point it should be a perfect circle where all lines(b-h-b) are equal to the radius. So instead of calculating Pi to infinite digits, there should have been an end. Remember, for a circle, the start point and end point are the same point. The measurement is fixed, not an endless series of numbers.

I know what you are thinking; if Pi is wrong, mister smart guy, what should it be AND why? Well, I worked backwards with the calculators to calculate a perfect circle, meaning that all leg lines and bisecting lines MUST be equal to the radius AND Pi MUST be divisible by numbers 1-9 or we will be back in an irrational situation. And Voila, Pi = 3.15!

And as a proof, here is Pi divided by all 9 numbers:

3.15	1	3.15
3.15	2	1.575
3.15	3	1.05
3.15	4	0.7875
3.15	5	0.63
3.15	6	0.525
3.15	7	0.45
3.15	8	0.39375
3.15	9	0.35

Imagine this chart by replacing 3.15 with 3.14159265 with a never ending series of numbers. Why did it never occur to anyone that something is wrong and let's find what's wrong. But NO! Let's just keep calculating Pi with smaller and smaller straight lines, even if the straight line length gets smaller than an atom. And to simplify this explanation, what I have really proposed is a calculator for a circle instead of straight lines trying to achieve a curved line. To be clear about the triangle calculators even with a circle divided into a billion triangles, it shows a 5 radius with a 4.9999999999 bisecting line. Simply put; if the line is correct at the 2 ends but off by a billionth in the middle, it is off 1/3 of the time. And since the circle is divided into a billion triangles, the circle calculation is off 1/3 of a billion times. Easy to see 3.14 becoming 3.15. As I have said this should be corrected to a 5 radius and a 5 bisecting line(height) then you have a perfect circle and 3.15 Pi. Now I have calculated the angles and line lengths to calculate circles, but my Union Contract says I can't tell you. But you have 3.14159 etc., and I give you 3.15, I think even a Physicist, oops did it again, a Mathematician can figure the difference. Pencils down, computers off, where's my raise?

BUT, BUT, BUT, Wait a minute. 3.15 seems so much bigger than 3.14+?? BUT If you subtract 3.14+ from 3.15, I hate math, but you will see the actual difference is approx. 0.008 around the entire circumference or approx.. .0002 per 1 degree! The human eye would not even perceive this change. HEY WAIT A MINUTE, you complain about no physical data(measurements) for circumference, but where is yours Mister Engineer? I have put the measurement procedures at the end of the book in the END NOTES. I would encourage everyone to measure Pi.

So to sum up Pi: Took Euclid, Archimedes, Pythagoras and current Pi to task for flaws, solved for circular Pi and only needed 2 decimal places: Had a good day! Now you have an Engineering answer for Pi

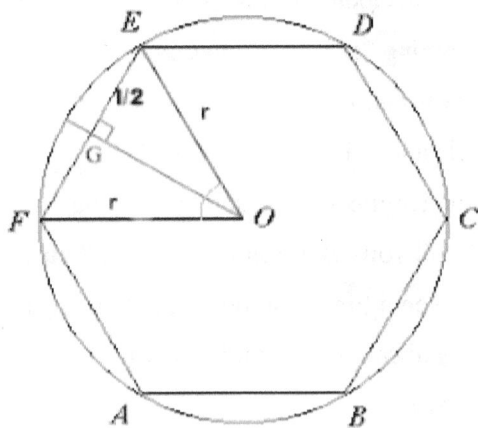

Sorry, got bored working on Pi and started reading about Entropy. Promise I will do a recap of vol. I next

ENTROPY

The second Law of Thermal Dynamics. Well, this is a fine mess

Another subject that Engineering really doesn't get involved in, but I see a flaw, so here I am. Let's just put Entropy aside for a moment and try to understand the Thermal Dynamics part. There are only 2 possibilities: 1. The system is closed and heat cannot escape or 2. The system is open and heat can go into or out of the system freely. Now this is very tricky because the Laws seem to imply they apply to a closed system as an open system could allow for some adding of heat or cold to the system As best that I can understand, Entropy is an offshoot of the second Law of Thermal Dynamics and energy transfer using statistical probability mathematics. Or more simply, over time the (Heat)energy of a system will tend to spread out evenly thru out the system(high Entropy). And during that process the particles(Micro States) may be in any position based on probability. But it refers to the complete system as the Universe

This is seemingly pretty straight forward but the outcome appears to actually be a Hot Mess. Since this is an Engineering View the very first question is:

1. What is the period of Time? A minute, a day, a week, a year or a life time? Engineers live for data and details. Nobody ever bothered to keep Time?

2. As an Engineer, the mechanics of a system is one thing that jumps out right away; What moves?, When?, Why?, How? What force is the spreading the particles out to equilibrium and high Entropy?

3. It is implied that this is a closed system, but unless the thermal shield can isolate the system from the outside surroundings(Air?,Universe?), it's heat will transfer out of the closed system and into the Universe. Therefore leaving behind energy-less particles trapped in a closed container. And then the particles are supposed to form into Micro-states? Do this experiment: get a pan of water, heat it to boiling, then turn off the heat and time how long it takes to cool. At the same time look up how many water molecules in 2 or 3 cups of water. Then divide the time by the number of molecules, this number is roughly the number of Microstates also. This is the approx. time each Microstate has to form and reform. Millions of Microstates would have to form at a blur, and also before the water evaporates into the Universe. Even in a vacuum, heat can be

radiated, just not conducted. The conclusion is there are no closed systems.

And finally, 4. What about Gravity? Even if the heat energy is radiated away from the particles, the particles still have mass and the four fundamental forces of nature are still available, although the strong, weak, and electromagnetic forces don't effect a single particle, Gravity can. And since Gravity, although very weak, has an effect over very long distances, it would be logical to assume that Gravity would begin pulling particles together. And before we dismiss this idea, this is exactly what happens every day on Earth. Earth's gravity causes tons of very fine dust (maybe energy-less particles?) to fall on Earth everyday. Starting with very small particles to build a Universe would take a very, very long time, but who knows, maybe this is the exact way this one started.

And since Engineers look at the mechanics of a system, it would fit nicely that heat is carried by electrons and electrons can carry lower and higher states of energy. Also electrons carry a negative charge that can repel each other and nicely explains the spreading out of particles and Entropy. But no, someone comes up with Micro-states and the random forming of arrangements of particles. How do these arrangements overcome the repulsion force? Just when it looked like this was going to make sense!

2nd Law of Thermal Dynamics and Entropy? Sorry, this Law doesn't seem very well thought thru. And since Engineers invented steam engines before the Laws of thermal Dynamics ever existed we should have just wrote down what we saw and did the Laws ourselves.

RECAP OF VOL I

SPACETIME – SO WRONG
Humans invented Time, How did it get into Space?

TIME
Only one master Clock, we all live by it.

TIME DILATION
Same story, if a clock doesn't agree with the Master Clock IT IS BROKEN!

THE AETHER – INTRO
Through out History people belived there was an Aether.

THE AETHER
Einstein dismissed the Aether but failed to explain how Gravity can distort empty Space and Bend Light. AND how can warp drives warp empty Space. Eather sounds right.

LIGHT CLOCKS
Supposedly explains Time Dialation, but since dialation does not match the master Clock, Light Clocks are just silly.

THE SPEED OF LIGHT
A structured Eather controls how fast Light can travel, but Mass objects can go as fast as they like.

GRAVITY

Newtonian Gravity allows for elliptical orbits, Eintein's Gravity distorting Space does not. Earth has a slightly Elliptical orbit, you decide who is right.

DARK ENERGY/DARKMATTER

My calculations show that Dark Energy/Matter can be accounted for if there is an Eather with minimum mass and energy.

INFLATION
An Eather can explain lots of things including Inflation. Negative charged Eather pulling positively chargedGalaxcies faster and faster.

THE TWIN PARADOX AND--- RELATIVITY
Poor Einstein never learned there is a Master clock so he created his own Reletive world. If you create your own reference frame anything can be right

OUR BIG BANG

First a little groundwork. What does an Engineering View actually mean? Engineers don't like to speculate much, we are data driven. When forced to guess about something, we still will gather as much data as we can, keeping the guessing to a minimum. Occam's Razor is our Mantra: The simplest Solution tends to be the correct Solution. And this means that the more things (complexity) that have to go right, the greater chances of Failure. Our Big Bang is simply the Big Bang that happened when you rewind the Expansion backwards to a time when the Universe was at a starting point, regardless of Theories of The Big Bang put forth by many People that the Universe was the size of a Pea or a grain sand or an atom at the Time of the Big Bang or that spontaneous particles or Energy caused the Universe to be formed. This is simply not true from an Engineering View. Here is the Simplest, easiest Solution that has some supporting data: No problem with running Expansion in reverse to the point where all mass In the Universe was in one area of Space, But not yet compressed by Gravity. Let's look at the compression and compression ratios. What is currently the most compressed or Dense Objects in the Universe? Black holes? Neutron Stars? Easily Neutron Stars, but why even ask? Because the real world Mechanics tells us that As the Universe compresses, there is a point At which spontaneous explosions can occur. This is called the Heilman Theory (just invented by me) and here are the details; Stars 8-15 times the mass of the Sun are capable of going Super Nova and Exploding, leaving a Neutron Star or a Black Hole. The Star runs out of fuel in the core and collapses and explodes. What is left is a very dense Neutron Star or a previously thought to be infinitely dense Black Hole. But we now believe that Black Holes radiate energy (Hawkins Radiation), so much for infinite Density. How does this apply to the Universe? The Theory says that if we compare the Density of the Star before the explosion to the Density of Star after the explosion, we can calculate a max. compression ratio possible for an explosion. NOTE: *Engineers do not like to talk in infinities. An infinitely dense Universe is simply not possible. There is a number, we just do not know what it is yet.* Now, we can apply this to the Universe. First, the Universe will have to be Compressed to give it the density of a massive Star. Then it will be compressed even further By the max. compression or Explosion Ratio. At that point the Universe will not only be Capable of exploding, but will by be likely to explode. So, what does the Heilman Theory tell us about The Big Bang? The most likely to occur will be that as the Universe compresses, the normal Star formation will take place and create a giant Star. This Star may include other existing Stars and will burn fuel (Hydrogen) very rapidly and create the conditions of collapse to causie a massive explosion. This Big Bang can occur, using the compession range of Stars that have gone Supernova, when the Universe is as big as a Galaxy. There is no breakdown of matter or forces, but background Radiation would occur at the time of explosion. What is very tidy about this Theory is that the numbers are real numbers used today for sizes and masses of Stars and the Universe. The Size of the Universe needed for Explosion (Big Bang) can be easily calculated.

But why is our Universe relatively flat? From this dispersal pattern, it is fairly obvious that The Big Bang was an off center

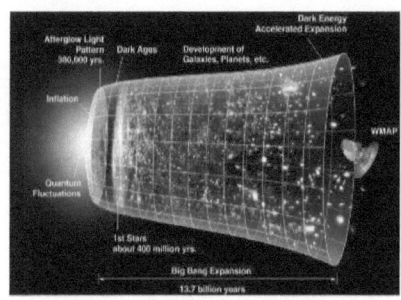 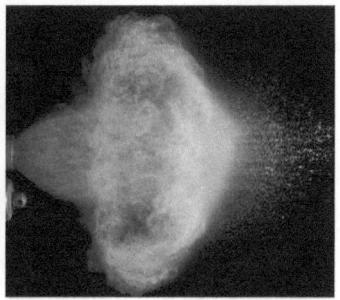

explosion. The left picture above is the distribution of the Universe, The right is a highly directional shot gun blast. Take away the smoke and very   similar

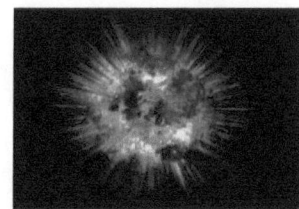 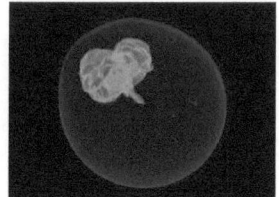

distributions.

Centered Blast   Offset Blast

So there you have it, very simple. This takes us back to Our Big
Bang. Maybe next book we will look at a prior Big Bang. I can
already predict that it will not involve a Universe compressed
down to the size of a marble or a Pea or an atom. There is no
evidence for that and at some point I am pretty sure $E=mc^2$ will
kick in and matter will convert to energy. We do see evidence of
that in Pulsars and Black Hole radiation. Bringing back
Common Sense.

THE DOUBLE SLIT EXPERIMENT

I am amazed by some things in Psyhics that

seems pretty basic, but never seems to be looked

at. What I am talking about is basically the Scientific Method,

but just common sense would do. Repeatability is the name of

the game in testing or experimenting. Trying to eliminate all

outside variables that could affect the results must be controlled

as precisely as possible. Along with that, a certain amount of

thought must accompany the procedure. In other words, could

something else produce the same results? The experiment was

ran with electrons, what happens if we run it with white light, a

certain color of light, a laser light, infrared, microwaves, radio

waves, etc. What are the dimensions of the wall with slits, size

of the slits, distance apart, how far is the detector from the slits?

Are the wall and detector exactly perpendicular to the emitter

(Gun barrel)? Is the target exactly halfway between the slits, is

the test ran with inert gas in the test area to prevent secondary

emissions from the air? How accurate is the Gun and what is the

repeatability? Is every shot from the emitter exactly the same? I

have reviewed many of these tests and have never seen a test

where all the variables are controlled. Especially the beam being

exactly the same every time, including being in phase with the

last shot. Every shot from the gun must be exactly identical to

the last for the results to be correct. This simply is not done.

Filters should used to block shots that do match exactly. And finally, why slits at all? Light can go thru a pin hole, slots just allow for more tolerance. My number one guess is that every experiment is ran in our atmosphere. Our atmosphere scatters light. Could this be the real cause of the anomalies. Is Quantum Mechanics even real. To better visualize this here are some pics to show the problems. Science already knows that an emitter can emit light in any direction. As proof, set a lamp in your living room, remove the shade and turn it on. You will see light go to every corner of the room. A ray needs to be controlled very precisely for the double slit experiment to be valid. To show this, here is a typical double slit experiment: Notice the pattern is very tall on the screen. If you are shooting electrons, the hits should be very tight. What this means is the Gun is not very accurate. Remember that even if you shoot just one electron it can go any direction. So how can we fix this? The simplest way is to use filters to only allow one beam to enter the slot. Here is an example of a filter:

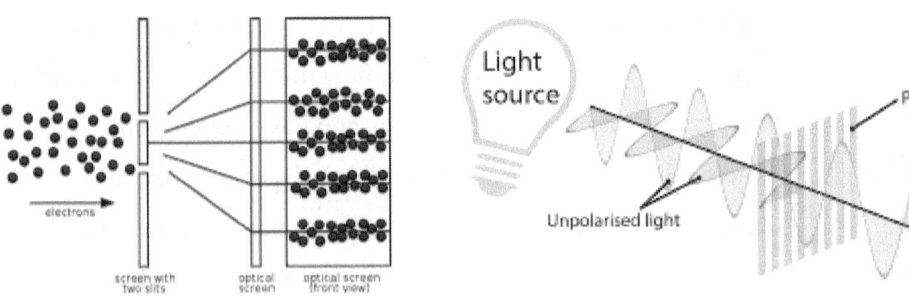

But what about up and down? That is what the pin hole vs a slot will take care of. Look how much more accurate we have made this experiment. So simple, but there is another thing that we must address, the fact that electrons coming from the Gun can go left or right. Remember the Scientific Methods demands that any other possibility for causing the same results be studied or eliminated. So we put a filter or wall on each side of the Gun to block very left or right shots. You see how this test could be made better? Finally, and lastly, there is the possibility that shots will not be in phase: The pic shows on top the wave we are looking for; the shot is identical every time. So, What does this mean? The current results are suspect. Run this experiment correctly, THEN claim victory and Quantum mechanics really exists. And to be totally fair, this is called an Experiment. An experiment is an investigation to learn something. An experiment does not have to be precise. But on the other hand only a Test can prove a Theory. Until someone runs a proper

test, all hypothesis in Quantum Mechanics are unproven and therefore Quantum Mechanics may not even exist.

Just an Engineering thought.

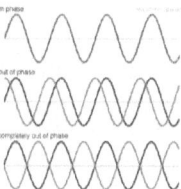

BLACK HOLES

Engineers are data driven and as I have stated before, we don't like to speculate on things. It either is or isn't, and if it is, show me the data. Now before I get off on a rant about unproven Theories being the basis for other Theories or Calculations being proof of reality, even if the logic seems reasonable, let's just say that Engineers want to see the data.

Black Holes seem to fall into the category of Seems logical, calculations seem to support them, but no real proof. Here is what the data or lack thereof, seems to suggest. The center of any object is generally viewed as the center of gravity. This is where most people and possibly Physicists go wrong. The center would have the least Gravity because it has no mass. Now objects can attract one another by their Gravity and eventually form a sphere, but the center will always have less. Gravity of course can push down on the core and increase density and gravity, but the center point will have none. I know the words infinity density and infinity Gravity get thrown around, but not true. What does that mean to Black Holes? The very Center of a Galaxy with no Gravity would have a very hard time forming a Black Hole. With all mass rotating around a center point, all with centrifugal force, how is a Black Hole going to form? Initially, it has no Gravity to pull objects to it and overcome centrifugal force. This where I think people get confused: Are there circumstances where a Black Holes could form? Yes, a supernova explosion could form a Black Hole, as well as a handful of other rare events. But this is like a million to one chance. So, as an Engineer, I see very little data to support Black Holes. Much less at the center of every Galaxy. And as a Note: Black Holes may not be Black because they do not let light escape, but because some of of the energy has been radiated away and the Black Hole absorbs any light coming to it. Pretty simple. And also along that line, the density and therefore gravity of a Black Hole does not appear to be all that great. If a Black Hole existed in a form like some Physicists claim, with high Density and Gravity, everything around it would be moving towards it. It looks like a slow dribble

To me. Which leads me to think of the possibility that what we think of as Black Holes are nothing more than a kind of Planet that can absorb light. I lean towards that conclusion because Planets do have some of the same characteristics. An event horizon where objects haveOnce difficulty escaping and massive Planets with massive Gravity appear to be just as capable of attracting matter. Once we figure this radiating of Energy we may see why Black Holes are just Black Planets.

THE (UN) CERTAINTY PRINCIPLE

Some of these Chapters I don't enjoy writing because the Principal seems fairly straight forward. Let's take an easy problem as an example. We want to know the position and momentum of a particle. The Uncertainty Priciple states you can only know one or the other because the act of measuring will change the relationship. Not always True. Suppose we have the case of the Double Slit Experiment. We setup up an electron gun to fire electrons, and then setup a detector to detect the electron when it strike the wall. The Gun will fire electrons one at a time. What we have now is the firing of the Gun starting the timer and striking the detector stopping the timer. We can now calculate not only the momentum with mass x velocity, but position any where on the route traveled by distance to detector / time. We now have a way to determine where a particle is at any given second. If a particle travels 2 feet in   ,2 seconds, after 1 sec from firing the particle is 1 foot from the Gun. We know the speed and mass, we now have momentum. WHAT IS UNCERTAIN. And to be certain of these results, we must have a Gun that will fire only one electron and know the distance to the detector presicely. Pretty simple, as is all of the Uncertainty Principal and Double Slit Experiment. And the main problem is that they are never ran with precision. Just think, even if we don't have the equipment or technology today to measure location and momentum, it will happen someday. So why would you build a hypothesis on an eventual obsolete Principal? This whole branch of Physics seems doomed to be obsolete.

FASTER THAN LIGHT

This Chapter is one of my favorites because I can use

Pure Engineering logic, unbiased by any Theories or assumed
Laws. It amuses me when some say that something can go faster
than light, BUT it doesn't break Eintein's Law that nothing can
go faster than light. I am an Engineer and as such I things going
faster than light. And based on what I see it appears Einstein
was wrong. Again, some people actually think Einstein said
literally that a nothing can go faster than light, meaning in empty
space. I don't like to use the word "stupid" but if the shoe fits!
Here are the big Six of things than can go faster than light:

1. Neutrinos – Only Theorized but could be
2. The Big Bang – The initial expansion Theorized faster
   than light.
3. Negative Matter – Warp drives. Still only Theory, but
   progress being made.
4. A light shined towards the sky – a beam of light waved
   back and forth. The beam would have to move faster
   than the speed of light to keep up. Although very real, I
   have not seen a test.
5. Quantum Entanglement at great distances. Pairs of
   particles moving simultaneously at great distances.
6. The Expansion of the Universe. – This is happening.
   Physicists qualify this fact by saying

That Space is expanding at the same time Galaxies are moving away from each other. Doesn't matter to an Engineer, How you are propelled is irrelevant, Speed is speed.

What is amusing about this list is that most Physicists qualify the statement to show that something goes faster than light, but they act like it is cheating so it really doesn't violate Einstein's Law. If Einstein and Special Relativity is wrong, at least have the courage to say so. I believe Einstein is wrong for a couple of reason's. Einstein began talking about the speed of Light or more correctly Electromagnetic waves and dismissing an Ether. But he continued on saying that mass objects could not come close to the speed of light. If you read any of Einstein's Theories, you quickly see that Einstein's Theories seem logical and are supported by calculations, but provide little or no details. Case in point, Einstein states that mass objects can't go faster than light. Why? Well as mass objects go faster, they become more massive and eventually it takes more energy to move more mass. Where did that come from? On Earth we have an atmosphere that provides resistance to movement thru it and yes it would take more and more energy to go faster and faster. But in the vacuum of Space, there is little to no resistance to movement. Einstein is wrong, there is nothing to prevent a spaceship for going significantly faster than light, maybe to infinity. An object will go as fast as an engine can push it. Einstein is probably right about Electromagnetic waves as they propagate all the same way(I happen to believe an Ether) and that method limits their speed. This is why when you shine a light from a moving train, the speed of the train will not be added to the speed of light. Space (Ether) will not allow it. But if you throw a ball (mass) from the moving train, the speed of the train and the speed of the throw will become additive. Mass objects do not propagate, but can be powered but an engine. Bigger engine, more speed. Speed is only limited by size of Engine and many times the speed of light is possible, theorictically up to infinity. But, but the Physicists like to say that as you approach the Speed of Light, mass objects get more massive, even infinitely so, requiring more power. No and No, And Yes, that works on Earth were we have an atmosphere to resist forward movement, but in the vacuum of Space, zero resistance. Bigger the Engine, the faster you go. Yes, even faster than Light.

NOTE: I mentioned that Einstein was great on Theory, but sadly lacking on details. Einstein proposed SpaceTime. He stated Space was 3D, But never said what it was made of. BUT WAIT, you say, Space is empty, it is not made of anything. Oh really, Einstein then states that mass objects can distort Space and that is how light can be bent passing a Sun.

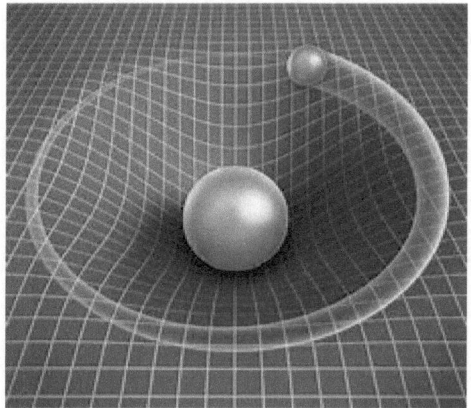

Well if Space is made of nothing, how can Gravity distort it? What is Gravity pulling on? This question could be answered by the existence of An Ether. But don't like an ether? Then put some QUANTUM particles in Space or Sub-Atomic particles. Something must be there for Einstein's Gravity to work or just admit Einstein's Gravity is wrong. I believe he is correct, but it needs an Ether to make 3D Space OR

Its back to Newton's Gravity. And finally, what is a warp drive warping if space is an empty void? CAN'T GO FASTER THAN LIGHT? Just another Theory that will look silly 100 years from now.

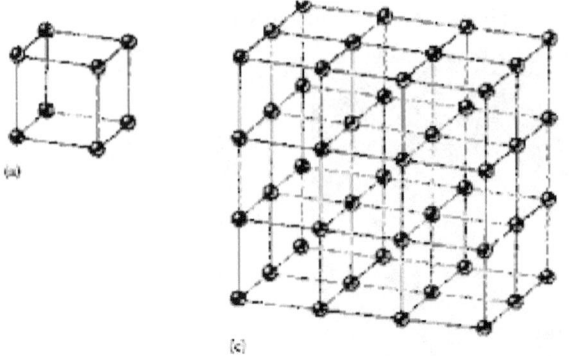

(a)

(c)

Possible lattices can make up Space.

FASTER THAN LIGHT STARSHIP

The last book I promised I would show you a way to go faster than Light. Engineers don't play around, so here it is:

As I was looking at the speed of Light capable frequncies, I saw the frequencies and wave heights defined. Something struck me as odd;

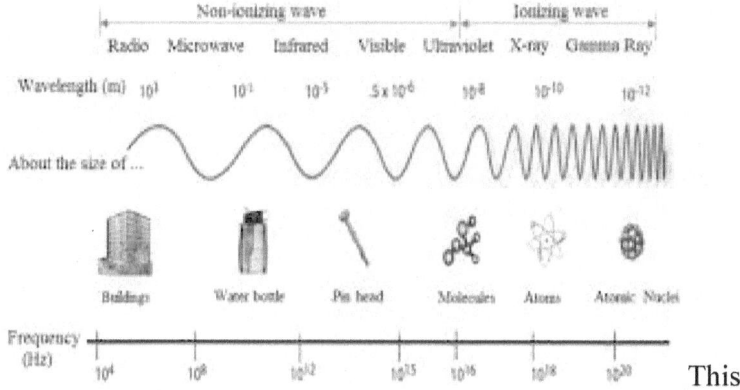 This

chart shows all the waves and The wave lengths and relative height of the wave. What this tells us is that the low frequency waves are taller and High frequency waves are very small. And what that means is that every frequency wave is carrying the same amount of energy. If a radio wave is 10ft long then 10ft of radio wave is carrying the same energy as 10ft of any other wave. 10ft

Of radio wave will equal 10ft of X-rays. 10ft of X-rays may contain maybe 500 peaks, but they very, very small. Radio waves will only contain 2 peaks, but they are very tall. So it is easy to see why all electromagnetic waves travel at the speed of light, they are all carrying exactly the same amount of energy. And to further explain, if you measured along the wave going up the peak and down the valley, the line distance following the wave is what is exactly the same. Ok, very interesting and very tidy, but what caught my eye was the chart below.

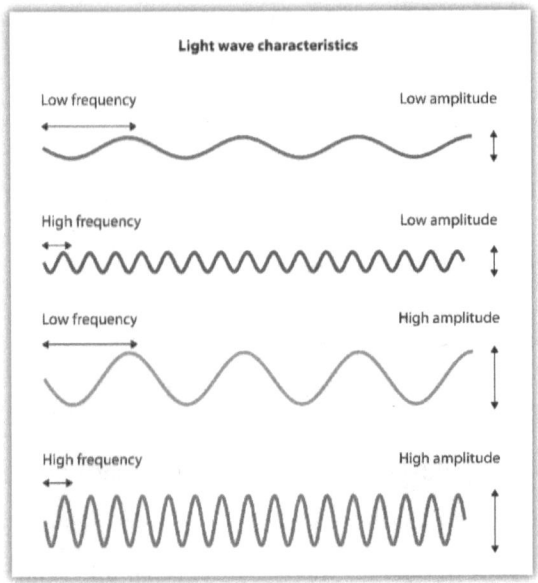

Physicists believe that high intensity waves carry more energy by increasing the amplitude or height of the wave. Since the frequency is fixed and the wave is taller, the electromechanical wave will have a longer distance to travel. Longer distance, slower wave speed to something less than light speed. In fact if you look at the pic above, the difference between low intensity vs high intensity may be in the 25% to 50% range. To be clear, the wave is traveling at the speed of Light, but the distance it must go is further. Now, you guessed it, if we reverse the process to low intensity(less energy), this will flatten (lower the height) of the wave. The electromagnetic wave will propagate at the speed of Light, but will have a shorter distance to cover. Meaning faster than Light! Let's take a ride in my new Starship, powered by an electromagnetic wave engine.

PHOTONS DON'T EXIST

I have already stated that Engineers are data driven and
In the absence of Physical data we will look for any data that
relates to the hypothesis, thereby keeping the speculating to a
minimum. For a period of time, it was thought that there was a
particle in a wave of light. This was because a certain amount of
energy was carried by the wave. This seemed to be much like an
electron or some mass particle. Well, along comes Cousin
Einstein, the nothing up my sleeve master, and all of a sudden
we have a bundle of energy (Quatum) that is carried by an
electromagnetic wave. But Einstein?, I thought you said that
mass objects can't achieve light speed. Oh, did I say that? Well
I guess that means this particle in light is a massless particle!
What?? But it carries energy, it must have mass. In fact that is
what 99.99999% of all mass objects do is carry some energy. So
Physicists respond with, it has both properties of a wave and a
particle. But an electron can travel in waveform and carry
energy, Why a massless particle? While a massless particle
solves the no mass object can go faster than light problem, it
creates other obvious problems. 1. You need mass to carry
energy. 2. You need mass to have momentum, $P=mv$. So guess
what Physicists did? You guessed it, they changed the Laws of
Physics to allow massless particle to carry energy and have
momentum. They actually claim that a massless particle can

have collisions with mass objects. How is that going to work? Imagine driving down the road and getting hit by a massless car. You probably wouldn't even know it!

So we have now reached the part that I least like about Physicists. Per the Scientific Method, for any Theory, you must list the other possible solutions and either test or plan a strategy. Not Physicists, they run with unproven Theories and even create whole branches of Physics to expand on unproven Theories (Quantum Mechanics, Special Relativity, General Relativity) OK, Mr. Smart Engineer, Just exactly what is wrong with Photons? Okkam's Razor, the simplest

answer tends to be the correct answer. What is the simplest answer that follows the Laws of Physics most closely? Wait for it, Einstein was wrong! Mass objects can go as fast, if not faster than light! Electromechanical waves are still limited to the Speed of Light, by the Ether I believe. But mass objects can blast away. This tidy's everything up. Light can have a mass object to carry energy, momentum is now correct P=mv, and the Light ray can have wave properties with particle properties. AND, because light now has a real particle in it, it can be bent by a Massive object. Yes, we don't need Einstein's gravity any longer and can go back to Newton's, which will allow for elliptical orbits. See how this is all coming together? And yes, you can still have the term Photon, it's just been redefined. But please drop the Quantum crap. Oh, I mean the Quantum term.

WHAT IS LIGHT

I hate to start every Chapter stating Physics is wrong, but a man's gotta do what a man's gotta do. The simplest answer tends to be the correct answer. Here is the simplest and correct answer: This chapter should be labelled as What is an electromagnetic Wave but as all electromagnetic wave have the same mechanics, so Light will do. As an Engineer, I look for as much data as I can find and draw conclusions from that. So here is the Truth, The whole Truth and Nothing but the Truth. 1. Electrons are an Energy carrier thru out the Universe.

2. Electrons can be free or can bond with Protons to form atoms, which also may have a Neutron bonded to the Protons.

3. All three are mass particles, with the Electron very small.

4. Stars burn Hydrogen which is 1 electron and 1 Proton. The concept: In a Star, atoms of Hydrogen are heated by fusion. The electron is orbiting the Proton closely in the first orbital position. As the electron absorbs heat energy, its orbit gets bigger and faster. This continues as it gets hotter and hotter. The bond between a positive Proton and a negative electron is very strong, but as the electron speeds up with more energy, it eventually reaches "Escape Velocity". The electron is thrown away from the Proton, wait for it, AT THE SPEED OF LIGHT which is the escape velocity. BUT, it gets better. The electron is moving along a vector, straight line. An interesting thing

happens to the electron. Because it is full of energy, if begins vibrating and the vibration is centered around the vector that it is travelling. A vibration that produces a high side of vector to low side of vector over and over. Because the electron is traveling at the speed of Light, the motion appears to be a wave, even though it is just an electron moving up and down. This cycle is called a frequency and since this electron came from Hydrogen, the frequency is that of light. At this point, the people who claim a duality don't know what they are talking about. And people who claim there is a massless particle in Light don't know what they are talking about. And people who claim mass objects can't go the speed of Light and beyond don't know what they are talking about. But wait there is even more; The same confused people will say this Light (Heilman)Theory is nice but what about Radio Waves or even X-Rays? Frequencies will be determined by the bonding techniques in atoms. In other words, copper has several layers(orbitals) of electrons. The electrons in the 4th orbital are held more loosely than an electron of Hydrogen. An electric current my be enough energy to cause a copper electron to fly off in Space at the speed of Light. But, of course, because the energy applied was less, the vibration or frequency will be less. And Radio frequency is way less than Light. So what about the higher frequencies? Just the opposite, higher energy applied to a tighter bonded atom that gives off an electron with a very high vibration rate. And as a NOTE: The Sun is overkill for producing one Light ray. That is why The Sun can light up the Solar System. And also, the Escape Velocity may be why Electromagnetic Waves cannot go faster than light.

This Chapter is in direct conflict with the Chapter Faster Than the Speed of Light. In that Chapter I show that if waves are the reason for the Speed of Light, then manipulating the wave shape will result in more speed. This represents the current accepted Theory that Light travels with waves. This Chapter is the Heilman Theory where Speed has nothing to do with Waves. Mass Particles are ejected from a Star or other sources of energy at the speed of Light. It doesn't matter what shape the particle is or the movement of the particle; a straight line, a wave or a jumping jack. It is all going the Speed of Light because that is the speed at which it was ejected. And yes, a vibrating particle would look like a wave at the Speed of Light, but movement of the particle would be the same whether it was moving 10 mph or 176,000 mps. And if this were true then it would also be true the particle could be propelled faster than Light if the propelling force was greater than the Escape Velocity. Even if you believe that the Escape Velocity cannot be increased, it doesn't change the fact that the particle could go faster than Light. And if you follow that logic to its conclusion, if a mass particle can go faster than the Speed of Light, a mass spaceship can as well.

Ok, I have talked a little about the mechanics of atoms and particles, but I feel I owe a more in depth explanation. Maybe next book I will give you the Unified Theory of the Universe to explain how and why things move. There are only 2 schools of thought: it is very simple or very complex. Since I am an Engineer, not a Physicist, maybe a couple particles will do. We will see.

NO SUCH THING AS WAVES

Up to this point I have tried to work with Physics to explain how things work. But now it is time to tell the truth, there is no such thing as waves. Of course this is just a matter of looking at the Universe a certain way, but the correct way. This is actually simple Relativity. Sounds funny, an Engineer explaining Relativity to Physicists, but since my ancestry is German, it's not that big of a stretch. To make this simple, imagine a young boy in a yard throwing a ball up about a foot high and then catches it as it comes down. Since the boy and the ball are moving laterally at the same speed Relative to each other, which is zero, the ball goes straight up and straight down. Now let's put the boy on a train moving 60 MPH. Again, the boy begins tossing the ball up and down. But because the boy and the ball and the train are all moving laterally at the exact same speed, to the boy the ball goes straight up and falls straight down. To a stationary observer outside, the train, boy and ball are all moving laterally at 60 MPH, but Relative to each other they are moving zero. This is simple Relativity. Such is the case with Electromagnetic particles. Yes, I said the Bad Word; particle. There is evidence that mass particles carry energy, such as Protons and electrons and others, so why not electromagnetic energy? The Wave perception comes from a vibrating particle, going straight up and down, but moving laterally at the speed of Light. If fact if you moved at the speed of Light you would not see a ray of Light, but just a particle vibrating up and down. The Wave perception is because it takes a particle a little time to go from full up to full down. And travelling at the speed of Light, the movement is perceived as a Wave. Waves are actually vibrations moving at speed, but this is not described as such and that leads to other misconceptions. It is easy to see that if you throw a stone into a lake it will displace water and form waves. But Space is not made of water and there is no Gravity. Throwing a stone at water in Space would look completely different. Objects in the Universe carry energy and make them have a natural frequency. And frequency is just cycles per time. Or, in other words, is one dependent on the other. Can you slow the speed of Light down and still have light? Yes, this happens all the time. Glass, water and atmosphere all slow down Light, yet the light can still be very intense. Can there be waves without Light. This is a little harder, but can be disproved rather easily. We know that light can only penetrate water so far. Simply put electromagnetic detectors beyond the point that the light doesn't shine. And there will no longer be waves. Waves don't carry electromagnetic energy, vibrating particles create a perception of waves. As viewed from a relatively stationary position. And to make this perfectly clear; A boy sitting in a train, tossing a ball straight up a foot in the air and then catching it will look exactly the same to everyone on the train whether the train is stopped, or moving at

60 MPH, or moving at the speed of Light because everything on the train will move at exactly the same speed as the train. But an outside observer may perceive events differently. As the ball goes up it will move down the tracks at a speed and as it goes down it will move down the tracks even further. This movement of the ball is perceived as a wave pattern. But which reality is correct? Ball straight up and down or Ball moving as a wave? Well, the key is that the boy must be catching the Ball and tossing it back up which is only accomplished by Ball straight up and down. The Ball could be moving laterally at the speed of Light, but so would the boy. Vertically, he is just tossing it straight up and down. And such is Light; A particle vibrating straight up and down, but moving laterally at the Speed of Light, giving a fixed observer the perception of a wave. Ride that Light Beam and just like the boy tossing a ball, all you will see is a particle going up and down. Relativity!

Ok, while we are at, this is also true for Light Clocks. For people who are not familiar with Light Clocks, this is a Theoretical Clock that helps explain Time Dialation. At rest, the base of the clock emits light vertically and at some distance the light hits a mirror and is reflected back down to the base. Besides an emitter in the base, there is also a detector. When the light is detected, it causes the clock to increase the Time by one increment. Very simple at rest, light straight up and reflected straight down; At speed, light straight up and straight down.

And since Light Clocks are just a visual aid, let's think about this; If your living room had a mirrored ceiling and you held a flashlight vertically the light would go straight up and reflect straight down. No matter how fast you ran around the room, as long as the flashlight was vertical, light straight up, reflected straight down. You just proved NO TIME DILATION and Light not traveling in waves, vibrating particle moving ho

APPENDICES

SCIENTIFIC METHOD AND FLAWS

Since the 17th century, the scientific method has been the gold standard for investigating the natural world. It is how scientists correctly arrive at new knowledge, and update their previous knowledge. It consists of systematic observation, measurement, experiment, and the formulation of questions or hypotheses. 1. Formulate Question/Hypothis 2. Define the Research Question 3. Review the Literature 4. Create a Hypothesis 2. Collect Data 2. Preparation: Make the Hypothesis Testable 3. Preparation: Design the Study 4. Conduct the Experiment or Observation 3. Test Hypothesis 2. Organize the Data 3. Analyze the Results 4. Check if the Results Support or disprove the Hypothesis 4. Conclusion 2. Look for Other Possible Explanations 3. Generalize to the Real World 4. Suggestions to Further Research --- ---------------------- I included this only this to readers how science is supposed to be done. I don't have A problem with Astronomers, or strangely enough Quantum/Particle Physics, they Started out as the wild bunch, but now seem to be on a quest for the truth, And other Physicists. The group that has gone astray, like Einstein, are the Theoretical Physicists. Relativity is a thought experiment. Special Relativity is a thought experiment. How does Thought experiments become

part of mainstream anything. Look at the Scientific Method.
Deep-field images from the Hubble Space Telescope suggest
there are 10 times more galaxies in the universe than scientists
previously thought, with about 2 trillion galaxies in total,
according to a study published in October 2016 in the journal
Science by Christopher Conselice, a professor of astrophysics at
the University of Nottingham in the U.K., and his colleagues.
[Video: Our Universe Has Trillions of Galaxies] About 100
million (or 10 to the eighth power) stars inhabit the average
galaxy, according to one of the best estimates, Conselice wrote
in an email to Live Science.

END NOTES

So as not to bore you with test data and procedures, this is the test procedure for Pi:

Ok, so here is exactly what I did and anyone can do it too. I bought a 100 ft tape measure. Now the way this works is that if we draw a 1 inch line and we could be off .1 the accuracy is 90%. But if we draw a 10 inch line and off by .1 we are 99% accurate And hence the 100 ft tape measuse, because 100 ft x 12 inches per foot = 1200 inches. So a 100 ft line drawn with the same drawing techniques as a 1 inch line, is 1200 times more accurate

Am I actually saying I will draw a 200 ft Dia. circle, then measure it? No, luckily no need to do that. Why? Because the curvature of a circle is the same for any segment. If you divide a circle into 1000 pieces, you only need to measure 1 piece and multiply that distance by 1000 to get the circumference of the circle or Pi. What does this have to do with my tape measure? Well, we only need to drill a small hole in the middle of the 1 inch line as a pivot, stretch the tape measure to any length you have room for, drill another hole at that point and attach a pencil or mechanical pencil to draw the arc. Create a pivot block by drilling a hole in some material(wood?), glue the loose drill bit into the hole, and when dry put the 1 inch end of the tape measure over the pivot pin. Now go to the pencil end, put a piece of paper under under the tape measure, put a pencil in the hole and gently move the pencil and tape in an arc around the pivot for about a foot. Remember the pivot block and paper must be held down securely, You now have an arc and all you need now is to determine how much of it you need. To do this you need a triangle calculator or formulas. The best for me is the Kesian Casio Isosceles triangle online calculator, 50 decimal places, but don't worry you won't need it. Set it to hypotenuse and base angle. Hypotenuse will be how far you stretched your tape measure to and base angle should be 89.5, which should be a 1 degree triangle. Mark this distance on your arc (this is a straight line distance). Measure point to point, but along the arc. Take that answer and multiply by 360. You have just solved

For Pi

www.ingramcontent.com/pod-product-compliance
Lightning Source LLC
Chambersburg PA
CBHW021043180526
45163CB00005B/2253